The Quanta Physics Theory

The Universe Reverse Vacuum Space Field Drive

e+h @ g 2.7c = FTL

TABLE OF CONTENTS

PREFACE

Rodney Kawecki born in Los Angeles California on September. 7,1953, has climbed the ladder of science and restored what he believes to be the final realm for the advancements of space travel. Having studied physics for more than twenty years Rodney Kawecki has leap-frogged into the mystery of the universe to achieve welcoming arms as his contribution to the earth science.

A SIMPLE EXPLANATION

The universe is expanding under the pressure of expanding space that is seeping up and between the galaxy's like water seeps up between two laying leafs in a lake from the bottom separating the leaves a part. But for the universe there exist a balloon type bubble at its center of mass slowly and at a constant the Hubble Constant growing from this concentrated round mass that was created at the same time by the big bang at the time of creation.

This bubble has put a time limit on our universe so a few physicists believe. Physicists have foretold a zero point time start for creation in all the physics books across the table and that this expansion rate seems to be evolving multiplying and erecting the space liquid. In the analogy, Rodney Kawecki has deciphered the strength and provisions this expansion is evolving at and simultaneously has discovered the resolution for this advancing super-gravity element. His predictions tell the story of an expanding universe that's time clock seems to be in correlation with Hubble Law. But also and most importantly it has made it possible for him to calculate conditions about the expansion of space that tell a different story than just universe expansion. It tells a story of what Kawecki has named as "The Universe Reverse-Vacuum Space Field Drive". It is a new theory that illustrates the expansion conditions of the universe and enabled him to create new equations not only for space expansion but also for a sound and special formula for a repulsive reverse vacuum space medium.

This new theory allows for faster than light space travel under standard conditions formed by an expanding universe the ability to travel distances across the cosmos in direct line with the ability to make sense of the material universe's effects of expansion with acceleration.

For the last of its kind, Rodney Kawecki has deciphered a new theory on space travel that takes us beyond the sci-fi stories of science fiction and time travel and brings us back to the reality and the standard model. His is a new theory blooming before your eyes that will take mankind on a final journey through the stars to places never before thought realized before. With these equations we will be able to advance space travel to places so far away in lesser time equated by any other physics theory that by only the gain of propulsion engineering and that capacity to do such for advancement that will not put us on hold by using the universe's own reverse-vacuum instrumental push force in addition with normal acceleration methods so we can travel to any newly mapped goldilocks planetary system within our grasps to extend new planetary inhabitances' for the earths human race. Relative light velocity theory and commons have been dis-in viewed.

Rodney Kawecki

CHAPTER ONE

The Universe's Reverse Vacuum Space Expansion

Reverse Vacuum Energy equation

$$e+h @ g\,2.7c = FTL$$

Dark repulsive energy is the name given to the mysterious force that is accelerating the rate at which our universe is expanding.

$$E = m^{\circ}g^2 \{2.7c\}$$

We have known that our universe is expanding that its galaxies and clusters of galaxies are moving away from each other at great velocities – since the observations of Edwin Hubble in the 1940s. It was thought that the rate of expansion would slow down over time as gravity gradually exerted a braking effect. However, about 20 years ago it was surprisingly discovered that not only is expansion not slowing down, it is actually speeding up. Some repulsive force is pulling the universe apart and this force was dubbed "dark energy".

Unlike mass, energy can produce an attractive or repulsive gravity depending on whether the energy pressure is positive or negative. The vacuum energy in theory has negative energy pressures; the problem with modern gravity theory is 'gravity' in space can be viewed as a physical action created by a physical thing called "The Expanding Void". An imagery of a round sphere embedded at the central mass of an expanding matter universe that is expanding into an idle space condition. Like any rare egg when cooked it separates by a force in its center. This is what the universe's 'Primeval Egg" experience shows us it is doing. Some thought in cosmology big thinking excites the idea that the universe retains no braking system to slow it down but more so and on the same *que'* that the big bang event is an event that is still in physical motion to this day and at its extreme and sows no recourse in slowing down now or in the future nothing will stop as it continues on its natural course and aggression or should it be stopped from day one.

Known in The Quanta Physics Theory, space expansion is dubbed from an expanding universe theory illustrated by Erwin Hubble's research shown as an expanding universe entailed into three parts. 1) Idle space. The condition of a reactive material or element that allows interaction of primeval forces. 2) The material universe. That part of the universe all made up of all matter existing throughout project universe and makes up of about 6 percent of everything. 3) The expansion void. A physical deity evolving mid-center the universe causing it to grow or expand. In biblical terms also known as the void.

Setting aside the factual basis of this expanding bubble as it has been called - the void is reverse vacuum compression of a fifth element. Amongst the five primeval elements it is space in a concentrated form caused by the explosion known to us as the big bang event. Its cause is the effect of an empty vacuum affect suppressed by reverse engineering formed in the middle aftermath of the creation.

This idea is equivalent to the cosmological constant used by Albert Einstein in his equations of general relativity and represents a cosmological constant and energetic response though above 'c' throughout space it is his notoriety of the "C" constant that was found to be invalid by the latter Hubble Observations. The idea of virtual particles per se in vacuum could also show the alignment of unknown territory in physics. But space can be considered to be made up in two parts. 1) Idle Space. That which we view daily in the sky. 2) the expanding element known to what is expanding aall galactic planetary matter throughout the cosmos.

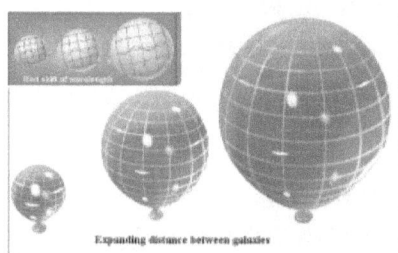

Expanding distance between galaxies

Repulsive Gravity

"In the theory of general relativity, we usually assume that the energy is greater than zero, at all times and everywhere in the universe," says Prof. Daniel Grumiller from the Institute for Theoretical Physics at the TU Wien (Vienna). This has a very important consequence for gravity: Energy is linked to mass via the formula $E=mc^2$. Negative energy would therefore also mean negative mass. Positive masses attract each other, but with a negative mass, gravity could suddenly become a repulsive force.

Even if the matter is somewhat more complicated than previously thought, energy cannot be obtained from nothing, even though it can become negative. New research results now place tight bounds on negativity, whereby connecting it with quintessential properties of quantum mechanics. If I went into space and had an empty box/container and went outside to fill that box with space, what would be inside the box?

1. Outer space, the void that exists between planets and stars, is not completely empty. Stars, planets, and moons keep their atmospheres by gravitational attraction. The density of atmospheric gas decreases with distance from the earth but there is no fine line where earth's atmosphere ends and outer space begins. If you take a box as far as possible from the influence of heavenly bodies and nebulae, you can theoretically say you are in "outer space" but outer space will still contain a low density of atomic particles (predominantly a plasma of hydrogen and helium), neutrinos, and dusts, as well as electromagnetic radiation, magnetic fields, and cosmic rays. All these elements are necessary to detect any type of gravity force. They are the elements.

2. Vacuum is space devoid of matter. It would be nice if we could say that a vacuum is space devoid of matter and energy but quantum field theory tells us we can't. There is always background energy called vacuum energy in a vacuum. The effects of vacuum energy have been experimentally observed in phenomena such as the Casimir effect and the Lamb shift, and vacuum energy is believed to influence the behavior of the universe on cosmological scales. Therefore, a vacuum is not empty space. It is the ground state of space with the energy of the ground state called zero-point energy or what's in The Quanta Physics Theory is called "Idle Space". The latter does not mean zero energy per se but the minimum energy below which a thermodynamic system can never go. The sources of vacuum energy are virtual particles or particle pairs that pop into existence and then disappear immediately such that they can't be directly observed. The reverse vacuum of space separates these known quantity's and keeps them separated helping in creating the attraction of matter throughout idle space. The repulsion vacuum is too great for the virtual particles steady detection vacuum energy especially of so small quantity.

For now, the only way physicists can analyze these virtual particles is through mathematical abstraction. Whether they have concrete physical existence is controversial. If you intend to fill a box with vacuum, you can only do this in theory. The vacuum "inside" the box (not the box itself) will have virtual particles and zero-point energy only but it also can contain a physical affect or reaction of unknown elements at the third quarter.

Reverse vacuum energy in theory has neither positive nor negative energies. It is the empty box. The problem with modern gravity theory is 'gravity' in space is repulsive and can be viewed as a physical action created by a physical thing called "The Expanding Primeval Egg". An imagery of a round sphere embedded at the central of mass of an all matter shelled universe that is expanding into an unknown yet idle condition of open empty space.

In the case of our universe this idle space action is an affect made by the big bang event itself. The false vacuum created by the big bang explosion expansion in an idle space condition that then evolves into model inflation; we discover a vacuum hole at the center of creation where everything exploded from leaving it empty it has been formed where the bang started. The big bang has done what it did with regenerating matter, it also dug a three-dimensional hole for a reverse-vacuum and frequenting empty space. It created a reverse vacuum centralize system for expansion by absorbing idle space flexibilities combining it separately as a whole at the vacant center and formed this void space element into a physical concentrated space at the center of fusion. A condition that illustrates why the universe is physically expanding by a physically embedded and concentrated body.

The Quanta Physics Theory

The figures are shown here:

93 billion light years (flat universe) astronomy dictionary Albert Einstein GRT

Round-about three-dimensional mass – a multiple of pi.

279,000,000,000 light years (^{2}c) pi

  3,000,000 light-years (minus a zero point for gravity) to maintain a substantial rate of velocity)

10,000 x 45 miles a sec (46.2 plus or minus 1.3 miles) = 450,000 miles a sec

Without using light frame measurement intervals"

(450,000 m/s) divided by L.S. = 372,000 m/s plus 78,000 m/s equals Speed without Limit

Equals: 44.7 miles per second every 3 million light-years

{450,000 m/s @ 99.999999 = 372,000 m/s} $E=mg^2$ (2.7$\square$)

  Will Quanta Physics theory physically prove to show quantum field theory wrong?

  A ship travels through idle space. It has the capacity to travel at close to the speed of light. Because of space expansion due to our expanding universe the ship travels 372,000 m/s. The Reverse Vacuum Effect caused by the flow of the universe's expansion allows zero mass ($m°$) receding vessels traveling through space be pushed away furthermore rather than be attracted by close planetary gravitation and so let's us travel through an idle space field at 99.999 percent capacity. This also includes faster than light velocities with no speed limitations because the reverse vacuum field gravity uses the equation $E=m°G^2$, instead of the latter $E=m°c^2$.

The reason is rather that gravitation in space is a reverse-vacuum natural dark matter push force, and rather hopelessly a physical opposite of the planetary attraction force.

With no exterior forces acting on our vessel except for the universes space expansion volume forcing it we are able to travel the full acceleration value of 99.999 percent velocity having nothing there to slow us down. Venturing into relativities infinite mass theories, we resume the value of the uncritical assumption of 'nothing can travel faster than light" due to the attraction field forces to the indifference principle with the expansion value that the ship that travels light speed due to the conditions of reverse vacuum response now travels twice that velocity based on the standard nature. What attraction theory does to slow things down the reverse vacuum response accelerates thing up faster.

As the capacity of acceleration rises due to the advancement of technologies our universe will continue to expand getting larger and larger over time. The idea that we will ever be caught up with god's creation is overwhelming. Today big "G' allows us to travel 450,000 miles in a single second of time but in the future this number will change. The idea that mankind reaches all the criteria in its engineering propulsion engines that will exceed this number relies on the idea that a reverse push vacuum expansion force allows an expansion of primeval accelerations can be measured with the foundation that the universe space field comes with its own advance acceleration condition that when or if reached will take more propulsive engineering advancement technology to oversee the greater velocity.

CHAPTER TWO

The Alcubierre warp drive

In 1994, Alcubierre proposed a method for changing the geometry of space by creating a wave that would cause the fabric of space ahead of a spacecraft to contract and the space behind it to expand. The ship would then ride this wave inside a region of flat space, known as a warp bubble, and would not move within this bubble but instead be carried along as the region itself moves due to the actions of the drive. It was thought to use too much negative energy until Harold Sonny White said that the amount of energy required could be reduced if the warp bubble were changed into a warp ring.

The Alcubierre drive, Alcubierre warp drive, or Alcubierre metric is a speculative idea based on a solution of Einstein's field equations in general relativity as proposed by Mexican theoretical physicist Miguel Alcubierre, by which a spacecraft could achieve apparent faster-than-light travel if a configurable energy-density field lower than that of vacuum (that is, negative mass) could be created.

Rather than exceeding the speed of light within a local reference frame, a spacecraft would traverse distances by contracting space in front of it and expanding space behind it, resulting in effective faster-than-light travel. Objects cannot accelerate to the speed of light within normal *spacetime*; instead, the Alcubierre drive shifts space around an object so that the object would arrive at its destination faster than light would in normal space without breaking any physical laws.

The theoretical physicist Miguel Alcubierre was born in Mexico City, where he lived until 1990 when he traveled to Cardiff in the UK to enter graduate school at the University of Wales. He received his PhD from that institution in 1993 for research in numerical general relativity, solving Einstein's gravitational equations with fast computers. He continues to work in this field, devising numerical techniques for describing the physics of orbiting black holes that spin down to collision.

Alcubierre published a remarkable paper which grew from his work in general relativity, the current "standard model" for space-time and gravitation. His paper describes a very unusual solution to Einstein's equations of general relativity, described in the title as a "warp drive", and in the abstract as "a modification of space time in a way that allows a space ship to travel at an arbitrarily large speed". In this Alternate View column, I want to explore Alcubierre's work and its implications.

There is also a second faster than light (FTL) prohibition supplied by special relativity. Suppose a device like the "ansible" of LeGuin and Card were discovered that permitted faster-than-light or instantaneous communication. Special relativity is based in the treatment of all reference frames (i.e., coordinate system moving at some constant velocity) with perfect even-handedness and democracy. Therefore, FTL communication is implicitly ruled out by special relativity because it could be used to perform "simultaneity tests" of the readings of separated clocks which would reveal the preferred or "true" reference frame of the universe. The existence of such a preferred frame is in conflict with special relativity.

General relativity treats special relativity as a restricted sub-theory that applies locally to any region of space sufficiently small that its curvature can be neglected. General relativity does not forbid faster-than-light travel or communication, but it does require that the local restrictions of special relativity must apply. In other words, light speed is the local speed limit, but the broader considerations of general relativity may provide an end-run way of circumventing this local statute. One example of this is a wormhole [see my AV columns in Analog, June-1989 and May-1990] connecting two widely separated locations in space, say five light-years apart. An object might take a few minutes to move with at low speed through the neck of a wormhole, observing the local speed-limit laws all the way. However, by transiting the wormhole the object has traveled five light years in a few minutes, producing an effective speed of a million times the velocity of light.

Another example of FTL in general relativity is the expansion of the universe itself. As the universe expands, new space is being created between any two separated objects. The objects may be at rest with respect to their local environment and with respect to the cosmic microwave background, but the distance between them may grow at a rate greater than the velocity of light. According to the standard model of cosmology, parts of the universe are receding from us at FTL speeds, and therefore are completely isolated from us. As the rate of expansion of the universe diminishes due to the pull of gravity, remote parts of the universe that have been out of light-speed contact with us since the Big Bang are coming over the light speed horizon and becoming newly visible to our region of the universe.

Alcubierre has proposed a way of beating the FTL speed limit that is somewhat like the expansion of the universe, but on a more local scale. He has developed a "metric" for general relativity, a mathematical representation of the curvature of space, which describes a region of flat space surrounded by a "warp" that propels it forward at any arbitrary velocity, including FTL speeds. Alcubierre's warp is constructed of hyperbolic tangent functions which create a very peculiar distortion of space at the edges of the flat-space volume. In effect, new space is rapidly being created (like an expanding universe) at the back side of the moving volume, and existing space is being annihilated (like a universe collapsing to a Big Crunch) at the front side of the moving volume. Thus, a space ship within the volume of the Alcubierre warp (and the volume itself) would be pushed forward by the expansion of space at its rear and the contraction of space in front. Here's a figure from Alcubierre's paper showing the curvature of space in the region of the travelling warp.

For those familiar with usual rules of special relativity, with its Lorentz contraction, mass increase, and time dilation, the Alcubierre warp metric has some rather peculiar aspects. Since a ship at the center of the moving volume of the metric is at rest with respect to locally flat space, there are no relativistic mass increase or time dilation effects. The on-board spaceship clock runs at the same speed as the clock of an external observer, and that observer will detect no increase in the mass of the moving ship, even when it travels at FTL speeds. Moreover, Alcubierre has shown that even when the ship is accelerating; it travels on a free-fall geodesic. In other words, a ship using the warp to accelerate and decelerate is always in free fall, and the crew would experience no acceleration gee-forces. Enormous tidal forces would be present near the edges of the flat-space volume because of the large space curvature there, but by suitable specification of the metric, these would be made very small within the volume occupied by the ship.

All of this, for those of us who would like to go to the stars without the annoying limitations imposed by special relativity, appears to be too good to be true. "What's the catch?" we ask. As it turns out, there are two "catches" in the Alcubierre warp drive scheme. The first is that, while his warp metric is a valid solution of Einstein's equations of general relativity, we have no idea how to produce such a distortion of space-time. Its implementation would require the imposition of radical curvature on extended regions of space. Within our present state of knowledge, the only way of producing curved space is by using mass, and the masses we have available for works of engineering lead to negligible space curvature. Moreover, even if we could do engineering with mini black holes (which have lots of curved space near their surfaces) it is not clear how an Alcubierre warp could be produced.

According to Miguel Alcubierre as light travels it sores laying itself close to the bubble fabric's surface. Like all mass sphere's they lay close to the bubble's surface. Since light has almost no mass it has no real effect traveling close to the bubble at constant acceleration. A mass though like a ship would affect the surface of the expanding bubble mass. This effect is called "The Alcubierre warp drive".

Alcubierre never took into his facts that space is a return vacuum. A reverse vacuum effect that pushes masses like planets, stars and even galaxy's simultaneously away from it. That its push force is greater than light speed that for now only acts a coordinate's factor in measuring acceleration. That it has no real meaning otherwise. And as a reverse vacuum space Field activity things move faster in space than they would if under the influence of an attractable gravity force like earths.

No in space things work way differently. Light sores traveling on top of the mobile plate that is expanding the universe making the material of its broken shell separate as it does. The universe's mobile plate that separates the bubble from the clusters and stars puts mobility to all the spheres that exist keeping them away from each other but what lays subtle are the stars and planets and solar systems inside their protective realm. Will they separate and become pushed a part after the expansion of the universe space shuttles and becomes idle? Will the interaction continue? Will the galaxies paralyze of their motion? If such an event were to happen the Hubble expansion rate will stop with it. And the universe will fall.

There will nothing left to keep it in mobility and everything we know about the universe will fall. The weight of all these things material and made of matter will idle out and all their weight will cause them to fall. Little by little or all at once the end will tell the story. All of the material mass will reunite into another singularity and again something will create a spark and it will all happen over and over again.

But referring back to reverse vacuum drive all material matter along with any mass and or energy is forced away from the existing space field. A ship traveling in such a cold atmosphere is simultaneously pushed from behind where it gradually gains its power of acceleration and expansion causes it to sore way beyond its subtle velocity. There exists little in its way as it gains into velocities far beyond its capacity. The ships capacity of acceleration is solely based on its engineering. Its engines ability to advance with propulsion that measures well with what space expansion will allow it too. So it seems Albert Einstein was right about one thing and that is that it is gravitation that makes the call to how fast anything can travel in the upmost of attraction or repulsive weight that steer's the universe.

CHAPTER THREE

The Time It Takes To Travel There"

Is time travel possible? Some think it is - but really is it? 4th dimension 'demeanor' measures by the threads of time as a dimension of itself. But space-time is based on the reality of the universes motion which is real and sit independent of itself being it is the universe itself. To say time is an actual dimension that interchanges within itself past, present and future actually means time is absolute and nothing in it can change. So time travel is impossible. (It either is(present) was(past) or never will be (future) because these aspects about reality and the universe are absolute. They are either present or they are not. Time travel would then have to come from the future earth and it doesn't exist at all except theoretically.

The question at hand is asked "What is Gravity"? The following question arrives "How does it work? To some persons of importance on this planet magistrate like Queen Elizabeth was very puzzles about trying to answer this question. In 1667 the queen assigned her imperialist Isaac Newton to research out answers to the mystery about gravitation. She asks Isaac "why do the apple fall to the ground'? The question should have read "Why do the apple fall from the tree" as a result for his task Isaac Newton discovered a mathematical sum that seemed consistent for answering the question about earth gravity (a truthful and correct answer should have read 'The apple falls because the apple has reached maturity raised from the planets natural elements – my queen).

Isaac measured the velocity and action of free falling objects. He discovered that no matter what the weight or size of an object mass – all masses seemed to free fall at 9.8 meters per second no matter what. The only indifference to this consistency of free fall he called 'attraction" or objects falling freely towards the earth's surface was the inconsistency of added force acting behind the free fall that even air diminishes away the higher you are.

No matter what – inconsistency was added by a push or acceleration overwhelming direction and pushing force behind an object in motion and changed the consistency of the objects free fall adding an additional pushing force to a freefalling object falling towards the planet's surface actually created a maneuvering passiveness to form by the freedom of the objects free-fall mobility. An object aviates by free fall passing through the earth's atmosphere measured only by the earth's atmospheric gravity dome or gravity ring. The part of the planet's atmosphere or bubble concealed by the planets momentum impressed due to the planetary orbital cycle mobilized within the fabric space grid plate.

As the earth rotates it is pressed physically against the fabric space grid. Because of the universe orbitration energy all galaxies as well as planets and star spheres embed themselves pressed and engraved against the space fabric itself. Because of the existence of matter formed by the big bang event that happens in all universe birth events throughout empty space the dark fabric by which the planetary matter laying relevant to this first beginning event is what generally mobilized a material space grid. Everything that exists in the universe was made in and from it thereby lies within its grid age.

While the universe was set in motion by the first big bang even impression of the first universe what followed was a good weightage of galactic and solar star system little bangs. All these following little big bangs all resulted by the same means as the first big bang in nature. The beginning big bang event started the grid motion in the dark element started by the inconsistency of what we might try and measure as immeasurability. While the massive first universal event mobilized an orbitrating cycle between matter and the dark element of space informing it into a grid smaller and shorter little bangs reformed the first event creating new and deeper grid plate age inside the newly formed universe. Inside these smaller galactic bangs even smaller and denser tiny bangs proceeded. Over the immeasurable time existence preceding us after the big bang event ordainments like Christmas lights sometimes proceed us outside are planets embedded space grid opening and allow us to observe what are called supernova events. Planetary stars and or possibly a system of stars or planets appearing from nowhere amongst the planetary observation we have already mapped of our solar system.

As you can see – the space grid gravitation element along with planetary gravity like here on earth measure and physically appear quite indifferent and they are.

Isaac Newton researched and deciphered the thread of object free fall passing through the earth's atmosphere at 9.8 meters per second. He also maintained that objects in free fall towards the earth's surface were an attraction. He led on to say that the attraction was formed by and objects mass energy that is weighed inside the molecular realm of all matter. He deciphered that in the same way a magnetic attracts other entities objects falling towards the earth's surface are attracted in the same way. It was later that Albert Einstein pursued the theory of gravity and assumed the idea that this same energy mass in dominate objects would built infinite mass in acceleration. That a moving object traveling at close to the speed of light could maneuver no faster due to the lack of energy. Relative to special and general relativity research – the speed of light was constant. Nothing could maneuver faster.

Quantum Superposition

Quantum entanglement is a physical phenomenon that occurs when particles such as photons, electrons, molecules as large as Bucky balls, and even small diamonds interact and then become separated, with the type of interaction such that each resulting member of a pair is properly described by the same quantum mechanical description (state), which is indefinite in terms of important factors such as position, momentum, spin, polarization, etc.

Quantum entanglement is a form of quantum superposition when a measurement is made and it causes one member of a pair to take on a definite value (e.g., clockwise spin), the other member of this entangled pair will at any subsequent time be found to have taken the appropriately correlated value (counterclockwise spin). Thus, there is a correlation between the results of measurements performed on entangled pairs, and this correlation is observed even though the entangled pair may have been separated by arbitrarily large distances. In quantum entanglement, part of the transfer happens instantaneously. Repeated experiments have verified that this works even when the measurements are performed more quickly than light could travel between the sites of measurement: there is no slower-than-light influence that can pass between the entangled particles. Recent experiments have shown that this transfer occurs at least 10,000 times faster than the speed of light, which does not remove the possibility of it being an instantaneous phenomenon, but only sets an active lower speed limit. (Quanta Physics Theory – "Instantaneous Space Travel Acceleration" read 'The Fabric of Space Gravity" 2013)

This behavior is consistent with quantum-mechanical theory, has been demonstrated experimentally, and is an area of extremely active research by the physics community. However there is some heated debate about whether a possible classical underlying mechanism could explain why this correlation occurs instantaneously even when the separation distance is large. The difference in opinion derives from espousal of various interpretations of quantum mechanics.

Research into quantum entanglement was initiated by a 1935 paper by Albert Einstein, Boris Podolsky, and Nathan Rosen describing the EPR paradox and several papers by Erwin Schrödinger shortly thereafter. Although these first studies focused on the counterintuitive properties of entanglement, with the aim of criticizing quantum mechanics, eventually entanglement was verified experimentally, and recognized as a valid, fundamental feature of quantum mechanics. The focus of the research has now changed to its utilization as a resource for communication and computation.

The counterintuitive predictions of quantum mechanics about strongly correlated systems were first discussed by Albert Einstein in 1935, in a joint paper with Boris Podolsky and Nathan Rosen. In this study, they formulated the EPR paradox (Einstein, Podolsky, Rosen paradox), a thought experiment that attempted to show that quantum mechanical theory was incomplete. They wrote: "We are thus forced to conclude that the quantum-mechanical description of physical reality given by wave functions is not complete."

Like Einstein, Schrödinger was dissatisfied with the concept of entanglement, because it seemed to violate the speed limit on the transmission of information implicit in the theory of relativity. Einstein later famously derided entanglement as "spukhafte Fernwirkung" or "spooky action at a distance."

It is by purpose and composition of this literature about the "f" force – of gravity and acceleration that the velocity of light is NOT constant. As you continue to read this book about earth gravitation and the universal space field grid which all material deities reside on – I explain a side about the gravity force which opens the passage wormholes not only to faster than light space travel but also – the only manner by which astronomical space field distances between planets and galaxies can be achieved. How time travel passing through the fourth dimension may not exist at all bending space as like a huge planet cannot even maneuver to how the speed of light may only exist with alignment to the density of the realm that may have to exist with it and why the space grid may not be pent ratable by any means. Being things to a more reasonable conclusion we should view the outer space grid formed as a rotating field grid which are planets, stars and galaxies reside on as they orbit and rotate on. Laying on top of Dirac's sea disk which acts as a grid mechanism operating as solar system, galaxies, Nebulas and stars systems the mechanism grids mobility relays independent to the embedded system of debris that reside its dominion.

Planetary matter occupying specific sectors on the space grid, above Dirac's sea grid sector lays a mist of open empty freedom space. Space that retains no ties to dominated existing occupied space grid mobilized by the existing matter residing its plate activity a cold accusable mist manifestation exist above the grid. Like quanta rays particles or light that only travels above the space grid activity the deep valleys and rivers that lay between the occupied planetary realms and mountains from which celestial deity's lay positioned throughout the universe , a space craft traveling at great propulsion speeds through the freedom dwelling that plays atop planetary Dirac Grid fabrications. With this disruption about the freedom space grid along came the reality of the ideas of time travel and short cut wormholes the idea that a direct slash into the Dirac grid plate would assure the ability to travel a shorter distance that following the mounted grid. A space ship like a boat called a submarine traveling below sea level – depending how deep a closer planet or star may exist traveling through the grid may have its advantages. Trying to manipulate the cold freedom cosmos field freedom where physicists talk about curving an atmospheric freedom atmosphere to create a maneuvering wormhole believing it will take particles sent into it backwards through time. The physical belief that there exist an extremely deep biological clock that can be manifested by a depth of velocity such existence could only be entered by intense overwhelming velocity shrinking space quantities fabric stringer fabric elements that allow manipulation of time. The same time and extreme density drama that allowed deep faster than light maneuver after the big bang event 13 billion years ago that bends and curved the dark space grid but not ripping it. It seems that anything made from a mother lode will not have the overwhelming strength to cause its breathier.

So the question about what gravity is still not really answered. A falling object descends towards the earth's surface at 9.8 meters per second. The earth elliptical passage crest line while the planet's atmosphere due to its oceans forms from it humidity water and rain. Clouds form above caught inside the planet's atmosphere with little means to escape due to the planets cycle continuum where thunder and lightning engage by the abundance of cloud vapor and electrical sparks layered into thunder bolts that stream towards a well vaporized surface point created by the presence of allusive sparks ascending throughout our planet's surface atmosphere. The earth rotates at 18.5 miles a second meaning that for every second of time that passes 18 and a half miles pass along the earth's equator.

All the elements I have just mentioned are what layer the atmosphere the lays between open empty space at the earth's gravity crest line and its ground surface. Due to the change in the seasons and atmosphere heat and cold vapors created on the opposing side of our planet at nighttime the different cycles our solar system orbits in its elliptical curvature around our sun that also creates different solar wavelengths heat and frost on the earth's surface great winds pass through the mountain terrain that a person can feel as a great invisible force of physical activity inside the atmosphere.

The biological turning of the planets journey around our sun passes measured in seconds compared to measuring the velocity of ground momentum we at the planet's surface measure in minutes and hours. Wheat we measure to be friction seems more so to be the activity due to the lack of constant force. One engages a propulsion force at the gas pedal of an automobile and calls it friction because the car slows down. Friction deemed as an electrical terminology holds little weight when it is actually weight that slows the automobile not any electrical drag. Objector force pushing against an extreme type opposing force measured in seconds on a clock seems more the reason a propelled machine would slow down relevant to orbital cycles measured by minutes band hours here on the planet's surface.

Planet gravity might be viewed in the same manner. A slow drifting object in orbital free fall is controlled by the dramatic element pressure force the planet creates as it orbits and rotates at extreme universal elliptical striving forces. The elements measure to be full atmosphere pressure a passive falling object renders through as it decelerates out of slow free fall. Each second the planet rotates 18.5 miles in its elliptical path – measured in the same measure of the second – the object falls 9.8 meters. As the second 18.5 miles the earth turns, the objects free fall velocity doubles to 19.6 meters and so on until the weight of the object free fall reaches terminal velocity. Terminal velocity meaning a velocity measured equal to the objects weight and the free fall of the object becomes terminate and will gain little to no more speed as it falls.

When we review and include the facts of the elements and pressure of the planet's atmosphere to explain gravity it's a lot different than just believing energy mass to be the source of our force of free fall. In quanta Physics Theory energy mass is what pods matter main chunks together that separate from main object core material – like pieces separated at the big bang 27 billion years ago. The center of mass holds the balance of allusive pieces existing separate from other chunks of matter and of different chemistry. Reasoning the facts and balance of terminal velocity energy plays no role to Newtonian attraction of a falling object – attraction is measured by positive and negative values and electrical terminologies having essence to magnetism. Einsteinium gravity steers to energy mass towards that attraction even through his celestial gravity terminology reframes a push and pull gravity force. Dominion stars create deep curvatures and bend space as the universe as a whole stretches the space fabric outwards towards the universe's outer rim. In this manner a balance occurs. More so – attraction in space is explained more as a free fall space curvature or stretching of space to a fabric that cannot rip when we research its matrix'.

The nature of his theory on gravity is formatted on the earth's natural elements air, earth, wind, heat and the nature that is enclosed on our planet's surface in space it is The Dark Element. Earlier physicists have defined the gravity force as an application of energy but a weak force which mass centralizes gravity as the center of mass in all objects. Even though it is the matter energy mass that everything is held together with at the molecular level – Kawecki believes it is not the cause of gravity itself. More so he entitles the gravity force as applied activity of the earth's natural surface elements and has explained the gravity force with the activity of nature itself rather than a weak force natured by energy mass.

Kawecki laterally opens the door to define the gravity force on the foundation of nature not energy and deciphers the equations and measurements that precede his theory. Why does falling objects no matter its initial weight or size all fall at a specific speed towards the earth's planet surface? What causes the action that precedes free fall and by what means does it happen? And finally – if the gravity force celestial or inner planetary atmosphere is not measured by the cause and effect of energy or energy mass that repels than what is it? In this book he will answer all these unknown questions about "The Path of Gravity".

When we talk about gravity we are not talking about the fundamentals of energy and mass but more so is the activity of zero point energy presence and surface weight. Space is different. We are talking about how similar or like material entities called celestials such as planets and stars repel forming the distance between them in space present on the fabric of space that impresses them into deep holes. In space yes it's the presence of planetary energy in the elements that commands the divide. But in the interior elements of a planet's atmosphere there exists the 'ladder of resistance' or an entanglement of free fall of an objects weight and acceleration as it falls. In the thinner upper atmosphere an object in free fall gains acceleration in layers through the atmosphere as the planet's atmosphere pressure increases making the object faster at variable speeds as it follows a curving free fall pattern towards the planet ground surface. Nothing travels straight down to the earth's surface.

The problem I have with ""G"' is in space similar matter (planets) according to physics repel and do not attract as today's modern physicists have been lead to believe over the years. Like a magnetic north pole and another north magnetic pole...they repel. According to the big bang theory if all was in advance a singularity bang universal, galactic or otherwise then these similar worlds of common ancestry matter ...repel... and don't attract wouldn't you agree ?

If this is the case than - a planetary gravity field "TORUS" otherwise...pushes away objects as in universe expansion theory explaining the pushing away expansion. Gravity thus on earth in this sense free falls at 9.8 meters per second compared with 18.5 miles per second it's rotational orbit...this is a free fall elemental falling object ...wouldn't you agree?

Do wormholes exist or can they be made using the energy capacity of a star? Even a star cannot penetrate the fabric planets and galaxy's themselves shelve themselves on in space. To think that breaking through the fabric of space would be possible lays the end of the wormhole travel short-cut theory in special relativity. It makes one believe whether or not Albert Einstein really believed that such things were special at all?

Rodney Kawecki tries to answer all these questions about gravity and more. This book will enlighten and supersede the most interested reader of advance cosmogony. Gravitation has at reach or range to infinity. However, it is the weakest of the fundamental forces. The gravitational strength is only $6*10-39$ of the strength of the strongest nuclear force.

Note: $10-39$ equals $1/1039$, where 1039 is 1 followed by 39 zeros. That is a very small number.

The strength of the gravitational force decreases as the square of the separation between two objects, as does the electromagnetic force. Although the gravitational force is much smaller than the other fundamental forces, it's impact concerns objects of large mass, such as planets and stars. Gravitation is what keeps the Earth and other planets in orbit around the Sun, as well as the Universe.

(Lift-Vector (repel) $+C^2=G^3$)3D.

CHAPTER FOUR

The Fabric of Space

Albert Einstein states space is a fabric and it is the curving of the space created by the weight of an object it sits in, that creates gravity. But if space is a fabric, shouldn't it have some measure of tensile strength? And if it does, the then the fabric of space itself is subject to entropy. There are parts of universe that are very dense, center of the galaxy for example. So naturally space is stretched downwards towards the center of the galaxy.

Einstein was right, again. Satellites that have been pulled slightly off their orbits show that the Earth is indeed twisting the fabric of space-time as it rotates, scientists said. They said their findings are the first to directly measure and prove an important aspect of Albert Einstein's General Theory of Relativity: that a rotating body warps and twists the "fabric" that combines the three dimensions of space and the fourth dimension of time.

"As the Earth turns, it is actually twisting space-time with it. Near Earth, the twisting is greater," "This twisting of space-time, which is also referred to as frame-dragging, has never been directly observed before, "Time and space, according to Einstein's theories of relativity, are woven together, forming a four-dimensional fabric called "space-time." The tremendous mass of Earth dimples this fabric, much like a heavy person sitting in the middle of a trampoline. Gravity, says Einstein, is simply the motion of objects following the curvaceous lines of the dimple.

According to Einstein, living in the universe is like living on a huge piece of soft elastic rubber. Space-time is a medium that has shape and form. Objects within this medium can flex and twist it. Every object in the universe pulls on the space around it, drawing the fabric of space-time toward its center. The more massive the object, the more it pulls. The amount of pull exerted by an object on the universal fabric is its gravitational force. So the apple falls to Earth because Earth has warped space-time in such a way that the apple must move toward Earth's center. More massive planets create a deeper warp, imparting a faster acceleration to objects that wander past.

Albert Einstein's theory of general relativity tells us that gravity is a curve in the fourth dimension of space and time -- and there's proof to back him up. What causes the curve is mass. Seriously weighty objects can bend the fabric of space-time. It explains why the planets orbit around the sun. The sun is so incredibly massive it essentially bends the space around it, pulling into orbit lesser objects (like planets) nearby. Similarly, with enough mass an object can even cause an otherwise straight beam of light to curve. In astronomy, that's called gravitational lensing.

Time is not immune to the effects of gravity either. It passes more quickly the less gravity there is, a phenomenon known as gravitational time. At the center of the Milky Way Galaxy seems like the last place to form a new planet, inhospitable and violent even. Stars crowd each other, whizzing through space like cars on a rush hour freeway while supernova explosions blast out shock waves and bathe the region in intense radiation. The very fabric of space is twisted and warped by a supermassive black hole's gravitational forces.

Geodesics: Universe gravity acts proportionate with each and every single planetary sphere and set of solar systematical sphere stars and galaxy's that in turn a fabric grid plate. According to Einstein a planet acts independent of itself as it rotates curving the surrounding space around it. The problem about Einstein physics is that celestial bodies don't attract they repel because they are formally of the same matter deity from forming out of the big bang. If this is the case, planetary bodies repel and don't attract. Accordingly a planet twists the present fabric space-time because it rotates not that the fabric makes it. It rotates due to the resistance between the planets and stars nearby that are also caught surrounding the dominion star our sun.

Because the planets, stars and galaxy's repel this is what causes them to twist. It is the weight of the celestial sphere that embeds the sphere in the fabric grid and it is the universe's spin that increase how deep in the space grid it retains. As the dominion star repels a planet within its grid it's pushed deeper into the space warp age. As the universe spins it is impressed deeper into the fabric. Maintaining a proportionate mechanism of spheres and planets twisting occurs because of the greater dominions grip on the planet pushing it into a rotational spin axis maintained by the universe itself and its motion. The planetary cycles of the spheres throughout maintain a constant magnitude and velocity due to the constant orbit and rotation of the universe. Twisting space is not independent to a single planetary action it is universal and is what we call 'universal gravity".

In November of 1919, at the age of 40, Albert Einstein became an overnight celebrity, thanks to a solar eclipse. An experiment had confirmed that light rays from distant stars were deflected by the gravity of the sun in just the amount he had predicted in his theory of gravity, general relativity. General relativity was the first major new theory of gravity since Isaac Newton's more than

250 years earlier.

Einstein became a hero, and the myth-building began. Headlines appeared in newspapers all over the world. On November 8, 1919, for example, the London Times had an article headlined: "The Revolution in Science/Einstein Versus Newton." Two days later, The New York Times' headlines read: "Lights All Askew in the Heavens/Men of Science More Or Less Ago over Results of Eclipse Observations/Einstein Theory Triumphs." The planet was exhausted from World War I, eager for some sign of humankind's nobility, and suddenly here was a modest scientific genius, seemingly interested only in pure intellectual pursuits.

The Essence of Gravity

What was general relativity? Einstein's earlier theory of time and space, special relativity, proposed that distance and time are not absolute. The ticking rate of a clock depends on the motion of the observer of that clock; likewise for the length of a "yardstick." Published in 1915, general relativity proposed that gravity, as well as motion, can affect the intervals of time and of space. The key idea of general relativity, called the equivalence principle, is that gravity pulling in one direction is completely equivalent to acceleration in the opposite direction. A car accelerating forwards feels just like sideways gravity pushing you back against your seat. An elevator accelerating upwards feels just like gravity pushing you into the floor.

The only problem Quanta Physics expert Rodney Kawecki has with general relativity is that the first car got there first? The understanding of gravity is that it is universal. Meaning that gravity on a planet experiences change due to the elements and pushing outwards of sphere energy in the essence of Einstein's planetary gravity. The bottom line is that planetary inference of its elements and energy change the density magnitude of the gravity force present from being universal by the adding of planetary forces acting within the gravity force field whereas space does not and it is universal in both instances. An object at rest experiences inertial because its acceleration is erected from its maneuver from being at rest state position or acceleration whereas in space there exists no external forces applying maneuverability except the fabric of the space grid itself.

Travel coordinates for the 14 light years Journey to the Nearest Star

The figures are shown here:

93 billion light years (flat universe) astronomy dictionary Albert Einstein GRT

Round-about three-dimensional mass – a multiple of pi.

279,000,000,000 light years (^{2}c) pi

3,000,000 light-years (minus a zero point for gravity) to maintain a substantial rate of velocity)

10,000 x 45 miles a sec (46.2 plus or minus 1.3 miles) = 450,000 miles a sec

Without using light frame measurement intervals"

(450,000 m/s) divided by L.S. = 372,000 m/s plus 78,000 m/s

equals Speed without Limit

Equals: 44.7 miles per second every 3 million light-years

{450,000 m/s @ 99.999999 = 372,000 m/s} $E=mg^2$ (2.7□)

Distance to Neatest Star

14 light years to nearest star = 8,189,475,840,000 miles away

One light year = 31,449,600 seconds = 584,962,560,000 miles

$E=mg^2$ @ 372,000 m/s

Today's modern calculation to the nearest star using relativity theory

Actual 7 year journey to nearest star

Days, months and years to neatest star

8,189,475,840,000 miles to nearest star

18,198,835 seconds @ g^2

303,314 minutes @ g^2

5055.2 hours @ g^2

210.6 days (to neatest star)

....

Using The Quanta Physics Theory in retrospect [a difference in 'G' matters is.{2 x L.S. plus .7c}

= 7 months (to neatest star) @ g^2

$$pE=mg^2(2.7c) \text{ and accelerating}$$

The universe has an expanding gravity force with a velocity equal to 372,000 m/s plus 78,000 m/s or to be more exact in numbers: 450,000 miles per second is the accelerating speed of gravity a carrier force that continues to grow greater over time. All this in accord with the common standard theory about the universe as we know it there are two different origins about the position of the universe. The first is its matter oscillates inside an expanding bubble formation. The second is that matter oscillates outside the inflation of growing pressure. We will be deciphering the universe oscillating outside the bubble whereas the standard theories about the universe all maintain their sufficient is operating inside the bubble formation. This indifference between the two universe theories of sufficiency and universe structures science have only researched using the inside bubble theory over the decades whereas Quanta Physics is the first theoretical science that describes the universe outside the inflating universe bubble pressure and expansion observing all the common laws of physics which render The Quanta Physics' Theory a new theory of advance physics.

We look at two parts about the universe little has been written about. The first is the earlier fate of its origin and the second where the speed of light stands between both of them. We look at the earliest avenue about the universe's origin and discover that in part of its collapse as a matured cosum the universe underwent a closed space barrier explosion. As part of its collapse the universe expanded in for the most part of itself to a degree it expanded beyond the density the universe is measured by today. As a closed eruption there existed no escape of the pressure that rose from the collapse. Because of the high degree of infinite pressure evolved and with breakage in the pressure built by the collapse and the pressure raised by it the strength of the expansion caused by the cosum collapse not only put everything in spin but also allowed only one way for the pressure to escape and that was to make concentrated. The young universe raised by density due to the raise in the pressure had no direction but to shrink backwards in on itself as it did when it collapsed. Nothing escaped the realm of the cosum bubble and its early mastoid.

Reviewing these two existing parts playing in the early universe's origin we for the second question have to answer the question about the speed of light. The speed of light seems to be an optical about the universe when it was first born to even today 13.8 billion years later in its history. The speed of light optical seems like it can't be reckoned with. Albert Einstein famous quote nothing can travel faster than light seems to useless to try and change its relativity. But the fact remains the speed of light at the earliest time of the universe's young origin during the expansion haven stretched and cause space to warp due to the flexibility maintained in relativity that classifies its character condensing of matter during this young time and birth of our universe space was not originally tough enough to withstand staying warped by the universe's creation. Space stretched and curved back on itself as relativity predicted in 1915 in general relativity theory says it did. To this effect space warped and shrunk back on itself allowing the physical nature of its flexibility as we know in today's modern physics. And by developing this way the speed of light haven condensed matter into the smallest of its extreme molecular structure formed. Everything we know as matter cannot travel faster than light. It is only that the warp age that space underwent doing creation that expanded and shrinkage that concentrated matter that space co-moving was warp enough and beyond its original fabrication allows for an open source speed limit.

What physicists do know about the wit of the universe and its creation that warp age of the vacuum fabric allows for a faster speed limit than that of co-moving planets, stars and even galaxies undergo even when they are set in motion by the wheel of inertia. From a subtle point and view space is motionless co-moving as a whole piece of the pie and everything in it is in motion due to an axis measurable by a rounded universal edge an edge whereas in a flat universe type doesn't exist. Event horizons as we observe them on the different planets like are planet earth are seen by the overview of the planets spin and is continuous. The universe shaped due to its existence inside a closed avenue of empty space was shaped with no means from the release of the built pressure created by its collapse nothing escapes nor was there any way for chemistry of its origin to escape out of the space barrier that surrounded it.

Everything in the origin of the universe was built by a greater force than what we perceive in the speed of light as a universal constant. The warp age that occurred during its creation is what allows the universe and all its matter to expand into as it is by the warp age that space is expanding from. Matter that reformed into a material mass that cannot escape the Einstein limit has nothing to do with the space fabric warped in the venture. Flexible to the degree a ship having the capacity to do so can travel faster than the celestial substance that doesn't attain a difference in its co-moving velocity.

The universe created from a single point as the standard theory explains it has been expanding and growing bigger since first sight of the anomaly. It is in this sense what allows that the galaxies are expanding away from each other and at a greater rate of speed than the velocity of light. As the universe expands the space expands along with it and according to the facts the space gravity is expanding faster in the forefront expanding edge we know as space.

It is space that even to this day and age of the universe that continues to fill in the cracks and curves it created in the beginning. By the weight the celestial planets, stars and galaxies tug boat through space by it is the space at the horizon end that is in expansion space that is moving faster than the matter that was formed in its way. From the single beginning origin in space the expanding universe widens the space between its galaxies that also widens the empty space between them. The further the object of the energy between two masses the least it is that any energy exist at all and is how travel at velocities faster than light can be weighed. (zg)

Hubble coordinates of the space foundation are the same that were created by Albert Einstein in the early twenty-first century. The factors for elevation in the modern physics science are led only by the acceptation in the rule. In a space field hence that is characterized as a repulsive carrier force there exist no acceptation that it is the force of space gravity that weakens the walls around the bubble that are universe is shaped by. It is the same force that created the cause and effect in the cosum's collapse. In a closed universe where all the matter existing in it is too small of an amount for celestial gravitation causes to be formed for a universe it might expand forever.

According to Albert Einstein the speed of light was rectified when the universe was smaller than the point at the tip of your finger. As the universe grew it became more and wider between the galaxies. The speed of light changed also everything we thought we knew about the universe was figured wrong. The gravity mass as we knew it grew a part over time and the energy it possessed weakened. The speed of light was no longer measured in the manner it was long before centuries earlier because the matter the mass energy generated from only existed in the dead of space.

CHAPTER FIVE

The Push of Space

In space, in outer space it is not all free fall coordinates that measures how fast a ship can travel at. It is divine intervention flight in zero point gravity and how the vacuum is. The deeper the area of the vacuum the faster a spacecraft might travel. There is most likely to exist deeper regions of space regions and planetary sections that free fall backed up by propulsion accelerations will push a ship to a most extreme velocity.

Over the decades the theory about gravity also researched by Johannes Kepler born in December 27, 1571 – November 15, 1630) was a German mathematician, astronomer and astrologer created a model of the solar system which he was held for treason for his idea. When we review those facts about gravitation at the celestial scale amongst the planets and stars we have to look at the whole picture. The universe spins slow dragging dark space with it as it rotates. As a whole it is the weight of the galactic matter that space forms into a rotating disk plate. It creates a virtual fabric that allows the heavenly bodies to become impressed into in deep impressions in the fabric. It's like the universes spin forms a non-penetrable surface relativity calls the fabric of space when in reality it is the universes spin that forms the surface grid. If the universe did not conform to this analogy disk dominion stars and suns would dominate space pulling in nearby spheres into its realm. But as it is the spin by the universe causes matter to push towards its outer rim edge. We look at the universe in allusive orbitration pulling the galaxies where in total motion it is the weight of the galactic matter that is slow dragged rippling the flat rigid space disk plate. Inside the galaxy's solar systems of planet hard ware retain an orbit due to the swift impression the suns retain on nearby planets but are contradicted by the universes spin. The density of a planet's atmosphere stretches into deeper realms in deeper regions inside the more dense galaxies that exist throughout space. But the universes spin is what causes the alignment of the planet structure that lay inside revolving galaxies, nebulas and clusters of stars.

If the universes planetary matter did not drag along its maneuvering the dark fabric the planets and stars making up the systems would fall downwards into the anchorage of great suns throughout the grid. It is the universes spin that the weight of matter traveling at terminal velocities some faster than others that create a balance in the universe all together.

On a subtle bases size mass of celestial spheres great and small heavy and gaseous are separated by 'similar' weight mass the spin of the universe creates, its energy mass only creates a shield for resistance amongst the bodies all matter is caught within the deep impressions of the dominate planetary realm we know as galactic celestials. Early physicists leaned down towards the idea that it was energy and attraction that the dominion of the gravity force manifested itself. But the truth of the matter is - like a moving car at is not friction that slows the carriage to stop but the lack of temporal force acting on it. Unless force acts as a continuum to moving objects like cars, boats etc.., they'll all slow from and cease to move.

Free fall in space and abroad earth surfaces venture gains by the lack of thrust. Terminal velocity acts as a balance between an objects 'weight' and its 'velocity'. As one free falls from the sky it is interrupted by the earth's rotation and tension of its orbit. Each second on the clock the earth moves 18.5 miles per second intervals per second a free falling object is pulled by the pressure elements (air, heat, wind rain) bestowed in the atmosphere becoming heavier and heavier as it reaches the surface. As it falls it's caught within the planets rotational track that layers the atmospheres invisible element forces twisting by the planets velocity. Bound by the planets elliptical path none of this change to any much of a degree over time, the second units of a time clock interval seems to be a constant amongst everything that exists throughout space even the pulsar energy star impulse waves.

Space seems to be divided by elliptical plates that each section of space is divided by. The solar system divides its space area from a proceeding existing star system or system of rotation formed by some nearby system of planets outside our system dominion of planets. Elliptical highways are formed between these plate activities throughout the grid. There exist many systems in a galaxy each rotating plate traveling faster or slower than the other. The elliptical path between them act as highway passages for spacecraft flight this discovery of elliptical plates was finally discovered in 2013 by a satellite passing outside our solar system.

There exist two layers of space that exist in empty space field. The zero point disk alignment the planets, stars and galaxy's lay and impress themselves in bending the fabric's upper disk plate or elliptical disk plate as I explained about in the earlier paragraphs. Then there's the empty space field grade lying above the grid. It are the planets and stars that layer the space grid with impressed wells that lay deep within the top of the disk plate. Solar winds form at the dividing caps of curved space and the planet disk plate each celestial implants with its positional force of its weight and rotation. Above the impressed planetary disk plates, like the one our sun has impressed on the fabric galaxies have impressed the greater impressed grid which 'open space' lays above them if the fabric of space acts with the impression of galactic matter than it is not empty because the planets and stars act in its midst. But above this darken element fabric lays an open terrain of empty space that acts as free space and is not occupied by celestial matters.

Free fall velocities and terminal speeds measure at all points and areas of un-occupied space. The free fall of universe galactic planetary matter as it orbits allowing momentum between the grid and empty zero point gravity (grid gravity) from which our solar systems rotate and orbit in and are formed caught inside deeper dense perimeters like in our Milky Way Galaxy.

History about the gravity force researched over the centuries have led scientist like Isaac Newton and Albert Einstein amongst only to mention a few to believe it is 'energy' or the 'attraction' of that energy inside matter that creates the gravity force. But is the impressed force of the universes action it takes amongst its celestial sphere and galaxies that it's layered zero point empty space element raises out from.

Neil Armstrong in 1963 headed an experiment on the moon by dropping a feather and a hammer at the same time as he sat sitting at his spacecraft's upper step form. He performed this experiment to challenge Einstein's theory about object mass and gravity. They both fell to the surface at the same speed. The hammer no faster than the feather but the fact remains that on the moon that retains as little as eleven per cent g-force that the lack of planet element existed on the moon's surface terrain. That eleven per cent of 1 g^2 comparison to earth's gravity equals about 1 per cent close to nothing. There existed no chemical elements as earth, air, wind or water on the moon. Its energy star content is near zero gravity as Armstrong could jump fifteen feet through the moons atmosphere only falling due to his measurable weight. What the experiment did prove was that the space that layered the moon's surface that retained no atmosphere at all – showed that space is a measurable zero point gravity realm even if it surrounds a desolate satellite like the moon.

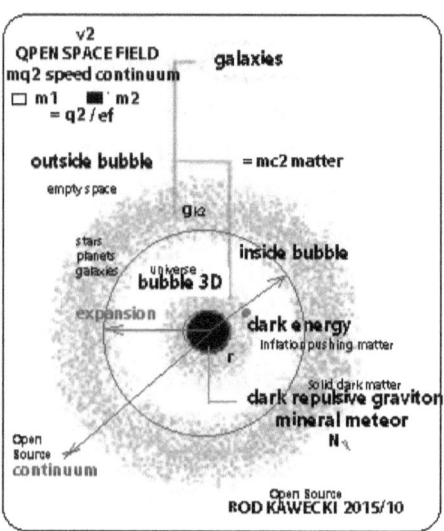

The Reverse Vacuum Gravity Expansion Space Field Drive

Reverse vacuum gravity is a review about space that allows a concentrated mass, like the balloon hypothesis about an expanding universe to supersede itself. Unlike a false vacuum the fate of the universe is not observed in the final hours. It slowly redeems itself backwards until the universe falls again. A time when all the forces we believe that hold the universe together diminish and are a collection at the bottom of the most gigantic gravity well emerges.

We view the universe under the scales made in relativity theory that measure the universes length at 93 billion light years in length. 93 billion light years is also the equation numbers for length of a mass under a gravity field meaning numbers to light speed are engaged by Einstein's light speed constant as well in his GRT and SRT theories but the manner in which he measured our universe.

It's up to us as entrepreneur to believe the universe can have validity to A.E's light constant verities of measuring everything about the universe a side from the universe being trapped in a bubble that is locked in all its behaviors at the velocity of light. Or shut the light out and agree that Einstein set us all up to a host and we should gallantly resume to the better deal.

Albert Einstein used light years instead of billions of miles for the length of the universe because light speed is the greatest length that was known to mankind at the time especially in a universe space field as great as the universe we reside in. It wasn't until Erwin Hubble whom discovered the universe was actually expanding that meta-parsecs lengths were invented to the equations that a longer length than light years were developed.

The Universe is still expanding and the big bang is still exploding it's retaining more and more area causing the galaxies to spread a part at their seams. How long will all this happen? Forever some believe? There is a limit some say. One of these persons was Albert Einstein. He believed the universe was *governed* by the velocity of light the fastest observation known to mankind. He believed that light redeemed everything we need to know about the universe. Though the speed of light is very fast - the universe is more than that. Besides his universe at the end of the ride is a big bang effect - we now know we can travel faster than light due to what's called "A reverse-vacuum expansion constant". That constant was discovered by Erwin Hubble in the year 1938. He is the one physicist that discovered the universe was expanding and measured how fast. Besides Albert Einstein he was the only one whom opens a new door to the cosmos the one doorway to the cosmos we use today.

But what is a reverse vacuum. It's a time in the duration of the big bang event that the speed warp-age and expansion of a primeval egg explodes and stretches across the cosmos from where it resided outwards into the maximum limit of what we know as existence. As all matter stretched across empty idle space the fabric of the space it evolved from and like a rubber band stretched out and stretched back into from point A back to point B amidst space.

It was this reverse vacuum activity that allowed an empty void to be formed in the middle of the action. When all matter particles and subatomic masses reformed back into gigantic chunks of matter debris at the middle it all formed a concentrated bubble of pressure and matter space that today is what expands the universe. Erwin Hubble explained it as a balloon that is now expanding and it seems indefinitely how and why what is written in the great novel about space.

Reverse Vacuum Space Concentrate

A concentrate is a form of substance which has had the majority of its base component (in the case of a liquid: the solvent) removed. Typically, this will be the removal of water from a solution or suspension, such as the removal of water from fruit juice. One benefit of producing a concentrate is that of a reduction in weight and volume for transportation, as the concentrate can be reconstituted at the time of usage by the addition of the solvent.

It will also allow the enlarging of the mass concentrate to increase if it subtle with a continuous flow of cold liquid such as water ignites with virtual energies its growth or expansion causing the solvent to multiple. The concept of expiration is related but legally distinct in some jurisdictions or the removal of water from a solution or suspension, such as the removal of water from a naked fruit of its juice. One benefit of producing a concentrate is that of space pressure reverse vacuum expansion.

Shelf-life depends on the degradation mechanism of the specific product. Most can be influenced by several factors: exposure to light, heat, and moisture, transmission of gases, mechanical stresses, and contamination by things such as micro-organisms the availability of a constant and abundance of supplies. Product quality is often mathematically modeled around a parameter (concentration of a chemical compound, a microbiological index, or moisture content).The universe existing of matter is only 5 % percent. It will keep its best quality for 12 to <u>49 billion years</u>.

With using these calculations we observe the expansion of concentrated space to last about <u>49 billion years</u>. (shown in above paragraph) We calculate the age of matter in our expanding space to be 13.5 billion years but we also account for how light travels through space and the equations in The Quanta Physics Theory that illustrate a universe expansion that is coordinated with an age of 40.5 billion years and since light no longer can act as its medium but was actually its stumbling block our universe has a 19.5 billion light years mile radius instead of 93,000 billion light year flat length.

But what does all this mean? The facts don't lie. Mankind has about 10 billion years left to evaluate its position in the medium. Whether it wants to stroll along as the universe slowly transpires. It decides whether to do something meaning venture the cosmos looking for hope or act as the most advance civilization in the cosmos and adventure to prolong all species throughout the cosmos in common worth.

The round or oval radius of the observable universe is therefore estimated to be about 46.5 billion light-years and its diameter about 28.5 giga par-secs (93 billion light-years, 8.8×1026 metres or 2.89×1027 feet).

Diameter: 8.8×1026 m (28.5 Gpc or 93 Gly)

Age: 13.799 ± 0.021 billion years

Density (of total energy): $9.9 \times 10-27$ kg/m3 *...in matter*

Contents: Ordinary (baryonic) matter (4.9%); D...

The observable Universe is, of course, much larger. According to current thinking it is about 93 billion light years in diameter. (Einstein's flat universe).

Light travels through space illustrating that the assumption of faraway places can be coordinated with the time it takes to arrive and actually catching up with the universe expansion using accelerations at faster than light processes. (refer to book " Is God Playing Catch-up with the Universe ")

In an expanding universe a ship has the ability to catch up with the universes expansion time variations by traveling at speeds solely relative to the universes space velocity. What this means is that a ship traveling at close to the gravity length of space expansion can travel to far away stars or galaxies that are expanding out of reach with time. We look at open space with the idea that the universe and all matter in it is expanding at a specific velocity. When observing tis time coordinates we discover that if we were to travel in the direction of the expansion we will be traveling towards the ends of time. But if we were to travel towards the sooner lengths of the expansion we would be traveling towards the early born time of the expansion thus towards a place in time when the expansion is hardly getting started. In other words we could travel in a direction normally known as the early beginning of the universes space expansion towards a point and time based in inflationary theory and expansion theory to be towards the big bang era.

We as a human species could have the ability to resurrect as a species that travels towards the early beginnings' of the universe. In other words as a species instead of idolizing on our planet waiting to a disastrous event that might happen according to history we travel towards the universes earliest beginnings era of time to advance as time travelers taking advantage of all obstacle's to our survival as a celestial species and before the end of times.

A Standard Universal Model

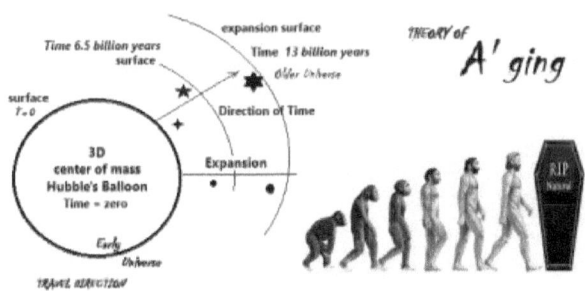

A ' GING